Writing Medicine

ISBN 978-1-329-20292-4

Contents

Foreword

Why is writing so hard? Who would think that in academia, of all places, where you're expected to write and publish, so many people would have such trouble with it? Yet many academics are afraid of the very thing that will ensure their professional success.

In fact, I was one of them. I could write just about anything, but put me down in front of a blank piece of paper or computer screen with the intention of writing something "academic," and everything inside me would freeze. My saboteur voice would take over and nothing was good enough.

Knowing how many faculty shared this problem, I kept looking for an answer both for myself and for the faculty I served through the Faculty and Academic Development Program at the University of Texas M.D. Anderson Cancer Center. For many years, I felt like a student of the writing process and hoped that one day I would break through and professional writing would become easy.

I don't know how I came across Joan Bolker's <u>Writing Your Dissertation in Fifteen Minutes a Day</u>, but when I did I found it revelatory. Joan had "midwifed" hundreds of Ph.D. theses in every imaginable subject. Was it possible that by writing for a manageable period of time every day you could complete a dissertation for Harvard or M.I.T.?

I tracked Joan down through her publisher and we arranged to meet in Washington, D.C. We had a wonderful, rich three-hour discussion covering a wide range of topics. It was thrilling not just because of her breadth of knowledge and her genuine passion for writing, but for the many things I discovered about her that kept me enthralled.

The first astonishing thing I learned was that Joan had been almost completely deaf for seven or eight years. Yet she was still a sought-

1

after clinical psychologist who worked with academics struggling with writing issues such as perfectionism and procrastination. In our own conversation, I experienced how deeply Joan attuned herself to the other person and how intuitive she was in sensing the real meaning behind words.

I discovered, too, that Joan and a colleague had established the first Harvard Writing Center in the early 1980s and that she had worked with faculty at Harvard, M.I.T., Drexel, Rutgers, the University of Massachusetts Boston, and many other academic institutions. Joan read widely and deeply, and loved to help people develop into competent, published writers. She knew what she was talking about.

Joan accepted my invitation to visit M.D. Anderson to talk to our faculty about writing. We enthusiastically accepted her suggestion that we sponsor a writing conference, which turned out to be a tremendous success. Joan gave the keynote address to an auditorium full to overflowing. She touched a nerve with our faculty, many of whom rushed up to the podium afterward with questions about their own writing stresses and defeats. You would never have known how profound her hearing deficit was as she spoke, listened, and answered questions with complete composure.

In the years following that first conference, Joan wrote a series of over 30 blog posts for us. She coached several of our faculty individually over email, and learned about the specific challenges of producing publishable work in academic medicine. Her blog posts were directly related to the issues she experienced our faculty dealing with.

I was surprised at how well one of her pieces of advice worked with clinicians and scientists, and that was "first make a mess, then clean it up." Getting started by doing freewriting turned out to be just the thing to get some of the faculty started on their writing projects.

This collection of Joan's blog posts is a gift to the broader academic community from a wise and generous writing coach (more like a writing shaman, I believe). In these tough times of reduced budgets and greater competition for fewer resources, writing effectively in journals, books, emails, blogs and even in documentation shared in multidisciplinary teams is essential to ensuring academic success. As Joan knows so well, learning to write more freely, more productively, and perhaps with less anxiety – maybe even with pleasure – is a goal worth striving for.

We hope this book will pass on to you some of what Joan gave us. It is an honor to have her as a mentor, coach, colleague and dear friend.

Janis Apted Yadiny, MLS
Associate Vice President
Faculty and Academic Development
The University of Texas M.D. Anderson Cancer Center
Houston, Texas

Preface

Before the publication of <u>Writing Your Dissertation in Fifteen Minutes a Day</u>, only my private writing clients had access to the advice and wisdom collected in that small paperback volume. Since its publication in 1998 it's sold over 100,000 copies, exponentially increasing the number of writers (academic and otherwise) who've used the strategies, advice, and encouragement it contains to help them write everything from doctoral dissertations to poetry. The book's reputation spread, and institutions as well as individuals recognized the value of the approach I offered. One result was an invitation from Jan Apted Yadiny, the director of the faculty development program at M.D. Anderson Cancer Center. Jan realized how valuable such guidance would be to her cohort of talented junior faculty, who were expected to churn out research papers and grant proposals while simultaneously running research labs and devoting themselves to patient care. Unsurprisingly, many of them struggled to get their writing done, despite its importance to their careers.

So Jan called me in, to provide a combination of in-person and online coaching – mostly the latter, via personal email and a series of blog posts for the group. As Jan notes in her preface, there's no way to share individualized suggestions and personal interactions after the fact; but there is a great deal of good advice, and even personal encouragement, contained in the posts. That's what you'll find here.

I want to thank Jan Yadiny and the writers at MDA for this collaboration. It was one of the best parts of my professional life.

Joan L. Bolker
Newton, Massachusetts
July 2015

1. Starting Out

This first post will introduce you to a new sort of writing process, and offer some observations about how to begin experimenting with it.

First, though, a mantra (mantras are valuable reminders when you're tackling hard writing tasks):

"Nulla dies sine linea" was written by Horace, in the first century BC. Translated into colloquial English it reads, "Write every day." What follows will show you how.

I know that you are pressed for time. Free-writing will help you use short periods of time (ten minute sessions) to get your writing started, strengthen your writing muscle, and keep writer's block at bay. It will also begin to prove to you something embodied in a second mantra (to be explained in the next post),

"Write in order to think, rather than thinking in order to write."

Here's how you do free-writing, a process first popularized by Peter Elbow, the author of Writing Without Teachers:

Sit down to write for ten minutes (on the computer, or by hand — but I suggest you start by writing by hand, because it's a somewhat different process) in a quiet place where you won't be interrupted. Begin by writing whatever comes to mind — don't sit and mull with an empty word balloon over your head, don't decide on a topic — just write. Don't worry if it's garbage; garbage is good at this stage. You don't need to write complete sentences, or spell correctly, or write about the same subject. You can write in whatever language suits you best at the moment; you can even write profanities, because NO ONE BUT YOU WILL SEE THIS WRITING. Just keep writing, don't stop, follow the pen, don't direct it (I know this may sound a bit like ouija board stuff, but there's a method to this madness).

If you find you have nothing to say, just write, "I have nothing to say," over and over, until something else comes to you (it might be, "Why am I doing this weird thing?"). I've never known anyone to be at a loss for other words after s/he's written "I have nothing to say" a few times.

At the end of ten minutes, stop writing. Then write, "Process note" after what you've already written and jot down a word, or a few, or a few sentences that describe how this writing felt to you. And now you're done — for now. (One of the questions you might want to consider for your process note is whether you know anything at the end of the writing that you didn't know when you began it. It's fine if you don't. But if you continue doing these ten minute free-writings the odds are excellent that you'll have such an experience.)

The following post will suggest a next step for focusing free-writing, but before I end this first one I want to mention a caveat: if you give yourself ten minutes a day for this sort of writing, every day, you may become addicted....

2. Second Thoughts

No, I'm not going to go back on what I said in the last post, but I want to elaborate on some of the ideas in it.

Let's look at freewriting again. Maybe you've tried it as often as you could over the past week or so. If it's beginning to feel useful and interesting to you, keep going, but try these next two steps:

• After you've written a quick, messy page or two, read back through it, and underline or mark anything that stands out for you. Then use a word, or sentence, or idea as the header for your next writing, and see where it takes you. This is the beginning of trying out iterative freewriting, a process that will gradually allow you to write in order to think.

• If, after you've done a page or so of freewriting, you're anxious about "not getting to the real work," try focusing the writing on a topic you want to think about, still writing quickly, without stopping, without fussing over getting the words perfect. Write another quick and messy page, but this time with a subject in mind.

• If you've hated the freewriting, and can't imagine ever getting used to it, work at writing in a mode you feel better doing, but still for a very short time, so that you can do it every day. (There's no point in sticking with something that you know doesn't suit who you are.)

• Clearly, the refrain is "Every Day." There are probably a few people on earth who can do wonderful, long pieces of writing in a stretch of several hours, once a week, but I haven't met any of them. If you decide, say, to postpone your grant or paper writing to the next vacation you have, you're likely to be anxious about it from now until then. Plus, you won't have a vacation if you keep that promise to yourself; you probably also won't have one if you don't, because you'll be busy feeling guilty.

• If you can establish a habit of writing every day for a short time, you'll have the chance to cook your ideas over time, continuously, so that when you do have, say, two hours in which to write you're much more likely to be able to accomplish something.

• Try "parking on the downhill slope" (I wish I'd invented this phrase — Ken Skier, who worked in the Writing Program at M.I.T., did): When you come to the end of your writing session, note quickly, at the end of what you've written, your thoughts about where you might go next with your hunches, or arguments or questions. This way, you won't have to start cold the next day; you'll have some prompts to get you back into your writing.

• Another important argument for working in very short stretches is B.F. Skinner's, that negative reinforcement extinguishes behavior. Translated into English, this means that if you keep punishing yourself about your writing ("Why can't I write for a longer time? Why can't I sit down and do the four hours of writing I said I would?") you'll eventually stop writing anything at all.

• Remember that even ten minutes of writing a day is an infinite improvement over none.

Each of the above has at its center the time problem. Many of you, I know, have very little time in which to write, but need to anyway. In the third post I'll mull some more not how to "make time," since that's impossible, but to use some of the bits of time you haven't considered usable before now. Freewriting is an easy way into doing this.

3. Some Thoughts on Writing and Thinking

What's the connection between writing and thinking? I was taught to write by thinking first, laying out my argument, and only then starting to write it all down. The trouble with this method is its potential for empty word balloons. B.F. Skinner advises writing in order "to discover what you have to say," that is, writing in order to think, not vice versa.

If you've had some success with freewriting so far, you'll know that your sense you may have had of "nothing to say" was misleading, that writing without trying to direct your thoughts sometimes leads to new connections, or discoveries, to the experience of "I didn't know I knew this."

(If you've given up on this sort of writing, though, because you're convinced that the think-first-then-write strategy works better, do consider carefully whether it really does, or you've chosen it because it's familiar, and you don't like change. There are some writers who can easily carry complex, ambiguous trains of thought around in their heads, and then just transfer them to written text, but this isn't a common talent.)

Establishing a daily writing habit that works for you, one that doesn't include interrupting the stream of writing to criticize what you've just said, will help make writing a place where you can try out ideas safely, discard more easily ones that don't work, and admit the natural contradictions, ambiguities, and wild speculations that come along with open-mindedness.

If you cultivate the habit of quick writing, more or less free (depending on what's worked for you in the past, and on your personality), you'll be able to do what Virginia Woolf called "capturing the diamonds in the dust heap." You might experiment with setting yourself a number of pages (messy ones) rather than a fixed amount of time, and seeing how quickly you can produce them.

Obviously, all of these methods call for serious revision of writing, which brings me to the next mantra, the invention of William G. Perry, Jr., for many years the Director of Harvard's Bureau of Study Counsel. Here's his description of the writing process:

"First you make a mess, then you clean it up."

What does "make a mess" mean? It can mean the kind of disorganized, self-contradictory, or outrageous stuff that might appear in your freewriting. It suggests that you carefully separate the creative process from the critical one, because combining them is a bit like driving a car with the emergency brake on: When you interrupt your early draft writing to go back and clean up anything from typos to contradictions it's very easy to lose your train of thought — you effectively halt the flow of ideas. Also, despite what your 4th grade teacher may have told you, most thought doesn't fall easily into standard outline form, and allowing for this in your writing will yield both better writing and more cogent thinking in the long run. If you note, say, in your lab notebook that you've just proposed two ideas that contradict each other, use this as an opportunity to write forward, and try to figure out whether this conflict tells you anything.

I've said "lab notebook," and it's not just my age that makes me visualize it as hard copy, and not notes on the computer. If you haven't tried writing by hand in a while, try it out for a few days or a week, and see, by keeping close track of how this shift works for you, what the difference is — in how you feel about writing, how much you get done, and what the product is like.

Computers are stunningly useful in producing writing, but not necessarily for early drafts. Oddly enough, I think they work less well for people who are very fast typists, that writing by hand may match the slower speed of thought processes better. Writing on the computer can turn out to be easier, faster, and much more superficial. It can also, especially in the revision stage, focus you more on tinkering with text, with fonts, with the machine, than on what you're trying to say. And it can confuse arguments wondrously quickly. (Oddly, one of the places where revising on the computer works best is in the revision of poetry, where there are few enough words to tinker with, and you can see whole texts side by side.)

You don't have to take this piece of advice, but I think it's one of the most useful ones I have to offer: except for very short pieces of writing, try revising on hard copy, double-spaced. You'll have a much better sense of where you are in the whole piece of work, and what sections need to be moved; you won't be tempted to scroll quickly through messy passages just because it's so easy to do. And when you're done you'll enjoy the lovely heft of paper in your hands.

4. Using, Not Making, Time

I want to go back to the discussion I promised in the last post, that I never got to.

People often ask me, "How do I make time for my writing?" It's clear to me that "finding time" is one of your most central concerns. Being fairly literal-minded, when I hear those phrases I note that neither is an enterprise likely to lead to success — when it comes right down to it, there are only 24 hours in a day. Sigh. How might we work with the time problem, then, of not enough time to produce the writing you want/need/have to?

The answer to this question is not in making, but in using time more realistically. Let's dispose first of imaginary time, the sort you invoke when you say, "I'll write this paper over Christmas vacation." All of you work very hard, most of you probably too many hours a week. There may be among you a few who live uncomplicated lives, without family or social or other obligations, or some who can get by indefinitely on four hours of sleep a night, or others who never need any down time to restore their energy. But most of you probably need vacation time for what it's labeled as. Imagining that you'll write the long overdue paper during that time as well isn't usually productive of much except guilt. (Later on, when I get to talking more about the behavioral principles that apply to writing, I'll explain why this result isn't particularly useful.)

How might you use time more realistically? First, as the above suggests, by giving up on the fantasy of long, uninterrupted stretches of writing time — they probably don't exist in your life. (A Boston doctor I know actually did "find" that sort of time, and write a book during a year when she took a "sabbatical" from her medical practice and joined her husband and family in Paris, where he had a fellowship — of course, she had to change practice groups as a result.)

15

Your job is to learn to write in much, much shorter stretches, the ten minutes a day I've written about already, the lunch hour (if it exists) during which you close your office door, and pretend not to be there, or, even better, go hide in a library where no one can see you, or phone you. If you and a colleague are both working on writing tasks, and have similar schedules, flee together to the nearest Starbucks or such, and hold each other to writing for half an hour, twice a week. Or do this briefly at the end of the workday (which involves giving up the illusion that once you get home you'll be able to write for two solid hours).

Use the ten minute stretches to make a quick list of thoughts you have for your grant proposal, or journal article. Consider a flow chart to map out your ideas, and put it on a whiteboard in your office, where you can add to it anytime an idea strikes you. If there seems to be a glitch in your argument, spend the ten minutes wrestling with it in writing; don't just go round and round in your head. If you're someone who thinks in the shower, get yourself a waterproof pad (they exist) and pen, and capture those ideas. Any time you're tempted to have a conversation with someone about a new idea, write it down first: ideas that are spoken before they're written down often evaporate into the air. Or, if you feel like you must talk about it, bring your laptop along, and write the ideas down as you talk.

If you're delivering a lecture, use the time on the plane on your way back to sketch out a draft of the paper it might become; don't set it aside for "when I have more time." Or tape record the lecture, and transcribe it as soon after you've given it as you can, noting the ideas that come as you write it. (The problem with not doing this is one I've often encountered with the academics I've counseled about writing: they have a substantial amount of potential writing tied up in talks they've given, but haven't managed to write up. Don't let your presentations grow cold before you get back to them: it's a waste of good material.)

Consider any ideas or experiences you may have had for using time that you might not previously have considered "real writing time."

A question I was asked recently is "What is the difference between writers who are primarily academics, writing dissertations, and those who are writing medicine, in medical settings?" Here's a very preliminary sketch of my answers, which are still in the category of guesses:

• Dissertation writers are probably more often in the position of writing on topics they've chosen because they're interested in them (from my point of view, this is the most powerful reason of all for writing a thesis).

• So, you could say that rather than being in the position of having to write x, they're more in the place of wanting, or choosing, to write y.

• Dissertation students more often feel a deep sense of ownership of their work, which tends to get judged fairly late in the process of producing it. I suspect your work, because it's often written on demand, and in a hurry, is much more difficult for the writer to own.

• Unlike you, many of the dissertation writers I've worked with have too much time on their hands, and have to struggle to structure it so they get their work done.

5. Mantras, Behavioral Principles, and Beginning Thoughts re: Revision

Let's begin with a few behavioral principles that can be really helpful in reshaping your writing process. B.F. Skinner described the difference between negative and positive reinforcement: the easiest way to say it is, the stick vs. the carrot. If any of you has raised a child, or a dog, you've probably learned that a bit of kindness and encouragement of good behavior works much better than harsh correction, not to speak of its being easier on both the participants in the educational process. Skinner demonstrated through experiments that negative reinforcement is also much less effective, and that positive reinforcement even works — in fact, may work better — when it's applied only sporadically, so called "random reinforcement." (My dog, whom I trained to come when called when she was running around loose in the park by giving her a biscuit, now comes every single time I call her — unlike lots of the other mutts — even though I don't always have dog biscuits in my pocket.)

What do dog biscuits (or their equivalent) have to do with writing? If you set your writing goals up so that you can meet them, that in itself will encourage you to write. If you also figure out, in the beginning, some other kinds of rewards that are particularly pleasing to you (taking time to go for a run, promising yourself a longed-for CD when the paper you're writing is done, actually getting to a movie after you've had a good week of writing), you'll probably find yourself getting to, and through, your writing work more easily.

If, on the other hand, you regularly chastise yourself for not getting enough done (no matter how much "enough" is), use each writing session as an excuse to drive yourself harder and harder, or promise to do tasks that really aren't feasible, and so, fail at them, you're likely to feel worse and worse about writing until, in Skinner's dramatic term, you "extinguish the behavior" — in plain English, you stop writing.

So it would seem more to your advantage, no matter how hard a taskmaster for yourself you tend to be, to focus on rewarding, rather than punishing yourself for writing, even if punishment is what you're used to!

An item that appeared every year on the psychology licensing exam when I was studying for it was the question about the Premack Principle, nicknamed "Grandma's Mashed Potatoes Law": No dessert until you've finished your mashed potatoes. In fancier language, this says that you can reinforce a behavior (like writing regularly) by following it with a behavior that pleases you more: so, if you're addicted to the morning newspaper, don't read it until after you've done your writing; if running, or stopping by your local coffee shop, is your pleasure, try to write first, and then make the run or the java your reward. This method, of course, is based in part on your having some control over your own time — which you may not have. But you can adjust accordingly (no glass of wine before dinner until after you've done your writing...).

About mantras: I've written a few of them in the earlier posts, like "First you make a mess (in writing), then you clean it up," or "No day without writing," or "Write first," or "A small bit of writing is much preferable to none," or….here are a few more:

"Don't talk about a new idea you've had before you write it down: Write first, talk after."

"Think carrot, not stick."

"Perfect is the enemy of good." (Donald Murray, a teacher of many writers, coined this phrase.)

"Write in order to think, not vice versa." (Skinner)

What I'd suggest you play with is making up your own writing mantras, remembering that there are as of now no personality transplants, so you need to take into account who you are. One of you might say, "Write first, clean later," another "I think I can, I think I can..." I had a succession of mantras while I was writing my last book, and some of them ended up as the prompts when I turned on my computer to start work.

About revision: I promise to get to this soon, but think it's worth your spending more time right now getting the messy part of the process down. Many of you are probably used to revising as you go. Try not doing this, but instead, let the writing take you where it wants to go. You'll have plenty of opportunity to revise later on. I promise.

6. Revision, The Critics, and a Question

I promised that I would talk about revision next. Even those of you who've had some success with "messy writing" may be worrying about how you're going to get from it to a finished piece that's accessible to, and useable by your readers.

Let's revisit for a moment the rationale for dividing the writing process into a messy first part, and a critical, cleaning up the mess second part. Earlier I described briefly the problem of "driving with the emergency brake on"; now I want to elaborate a bit.

Composing texts and revising them engage very different parts of our thinking. When we compose or invent texts we allow our imaginations to wander, consider different paths, perhaps even contradictory ideas, without judging them, just put them down on paper or screen. The richness that can result from this constitutes the "mess." When we revise, we say, "OK, playing is over, it's time to figure out what's here, discard the nonsense, organize the rest, clean it all up, and see what I have here that's useful." If you only do the first, messy part, you won't end up with something anyone else will understand or want to read; if you only do the second, critical half (this sometimes takes the form of writing to try to get every word and thought right the first time) you're much more likely to produce something dry and uninteresting, and you'll suffer a lot while you do it.

Not all of the writing process is fun (this isn't news to you), but allowing yourself the pleasure of holding off criticism for the sake of freeing up your mind to play with, invent, and explore ideas you hadn't imagined when you began, is much more likely to bring you back to your desk eager to write, much more likely to produce deeper and richer written work in the end.

Today let's look at the beginning of the revision process. When you start scrutinizing the pages you've written, don't expect a single run-through to get you a complete paper. Revision that allows several iterations is both easier to do, and more effective. Think of the process as asymptotic, where your aim is to get closer and closer to the asymptote that represents perfection, even while you understand that perfection isn't possible.

Here's how you begin. Have at hand double-spaced hard copy of your messy draft. This is worth trying even if you've always revised on the computer (if after you've tried this new way you want to go back to the computer, that's fine, but give this method a serious attempt first). There are several reasons for using hard copy:

• You have access to all of your text at once, without fiddling with the computer;

• You'll be less tempted to begin manipulating the text, moving passages around in a rush;

• You will be much better able to see the shape of your text;

• You won't be tempted to substitute spell-check and grammar-check for real revision;

• Your original will remain intact, and represent a stage of your paper that's important;

• It will also be there if you change your mind about changes you thought would work. (I know that you'll always have a copy of the original in your files, but making comparisons between versions is much more of nuisance);

• This is perhaps a personal idiosyncrasy: some of us enjoy the chance to work literally *on* the written text.

Begin the revision process knowing that the first run through isn't the last, so you don't have to catch every problem. Read as quickly as you like, marking things that catch your eye either as problematic, or as interesting thoughts to develop.

You can decide in advance what you're going to focus on in each successive run through (organization, development of ideas, phrasing, etc.), or you can give yourself the freedom to just notice what you notice. Make notes for yourself during each iteration, either in the margin of the text, or on a separate pad. And don't, whatever you do, scratch out something so that you obliterate it: you can't be sure that you won't want to resuscitate it later on!

Repeat this process as many times as you can stand, or until you feel that you've caught most of the problems in your draft. Now you can enter your corrections into the computer, and print out a copy that includes them all.

After you've worked your way through this revision process, put the paper aside for a while (depending, obviously, on what sort of time pressure you're under to finish it), and let it (and yourself) rest. When you feel sufficiently recovered, read through your revised text, and ask yourself, "Is this really finished, or do I just want it to be?"

In my next post I'll offer some suggestions for more specific kinds of revision that are part of the process I've described above.

Once you've tried it out, reflect on how this method has (or hasn't) worked for you.

Here's another issue: Many of you are thoroughly bilingual, but write in English. I've been asked if it's OK, and useful, to sometimes do the early drafts of papers in your native language, unless you're dealing with data sets, or literature reviews, which are probably too cumbersome to translate back and forth. Once again, I'd urge you to experiment: Try writing an early draft in your native language, and see if it's easier and more fluid for you. (If all of your education in your field has been in English, this may not be practical.) I can imagine, for example, writing about your hunches, your concerns, and your theories, or perhaps your abstract of your argument, in your first language. The interesting question is, what will happen when you then try to translate this work into English? Maybe it will be too cumbersome a task. But it's possible that the process of translating will give you more insight into what you're trying to say. No two languages I've ever studied captured ideas in quite the same way — even as a beginning student, meanings and resonances of ideas I was translating from German were broadened for me as I thought about what the closest way of replicating them in English was. And if writing in your first language makes writing easier, that's a pretty strong argument for doing it!

7. The Critics, and More Revising Hints

First, let's talk about the critics of your writing, about the times when you will get harsh feedback, and how difficult it is to absorb and use. How to think about it?

Strong negative feedback isn't surprising in a culture in which the stakes, and the expectations of your work, are so high, but you need still to be able to write even though the feedback is hard to stomach. How to do this?

My fantasy would be to retrain critics to deliver their suggestions with full awareness that the text was written by a person who has both a stake in it, and feelings about it. But this is fantasy, not a realistic goal. How might you put aside the anger and/or depression engendered by harsh criticism, and use whatever in it is correct and helpful?

Assume that any piece of writing, no matter how talented its author, can probably use an outside opinion before it goes to a larger audience, and that your editor, while perhaps a mite deaf to his/her own tone, wishes to be helpful.

Remember that this is one reader's opinion, and that other readers might not agree with it (and that if a paper is groundbreaking it's more likely than not to be controversial).

Remember that the work ultimately belongs to you. A journal or granting agency can decide not to publish or fund it, but if you're asked to change your work in a way that seems wrong to you, you don't have to. Find another journal, or another agency.

Don't invite overwhelming criticism, particularly of an early draft, by asking a first reader to "tell me everything that's wrong with it": there is no need to be quite that macho, particularly if the response stops you from writing!

Another, more useful way of approaching a reader is to formulate some questions you'd like answered about your draft, e.g., "Does this feel complete to you? Do the data sufficiently support my argument? Can you follow my argument easily? What stood out for you most in the paper?" — or any other specific questions for which answers would be useful. Don't ask someone to give you a global evaluation, because there's not much you can do with one!

Search among your colleagues for someone whose acumen and sense as a reader you trust, and then see what you can offer in exchange for his/her invaluable editorial suggestions.

No matter how smart, hardworking, and nice you are, not everyone is going to love you, or your writing.

Remember how unlikely it is that even your most brilliant writing is perfect.

More About Revision:

Use freewriting to help you revise, but use it a bit differently:

In five to ten minute spurts of writing ask yourself, over and over, "What is it exactly that I'm trying to say in this intro/paragraph/conclusion/paper?"

More broadly, use your time to write about focusing your early draft into a real one, asking yourself questions like, "What have I said so far? Is it true? If not, why not? What essential points have I left out? What's the most important point that I'm making? If I were the reader, would I be convinced by this argument?"

Try writing a quick abstract of your zero draft, to see if it's possible to. You may find there aren't enough ideas yet, or that you have the makings of three different abstracts.

Even while you work to correct, modify, and pare down the messy writing to a neater, clearer, and more succinct draft, try to keep your mind open, by continuing to ask yourself questions, and, as the literary criticism folks would say, "interrogating the text": If this assertion isn't true, what might be? Would I really swear by these results? Can I stand behind them? What's my wildest guess about these data? (You'll discover, as you practice this skill, that it gets less and less worrisome to ask yourself such questions — and that you become more comfortable answering them, particularly since some of the answers will be quite useful.)

More about revising next time.

8. Some Questions, the Details of Revision, and Some Mantras

Begin with some questions about this experiment you're engaged in:

• How are you managing time for writing? Which of my suggestions are useful, which not? Have you found some useful ways of your own for dealing with the issue of not enough time for writing, and if so, what are they?

• Have you considered who among your colleagues could be the right sort of first reader for you? Have you talked with that person yet?

• Which suggestions for revising have you tried, which have worked, and which haven't? Do you have some ways of revising that you find particularly useful?

• If you are bilingual: How do you think this affects your writing? Which language do you write in, at which stage of your project? How would you describe the difference between writing in your first language, and your second?

To reiterate some important mantras, and add some new ones:

"Nulla dies sine linea" (Horace) — loosely translated, "Write every day."

"Perfect is the enemy of good." (Donald Murray)

"Write to discover what you have to say." (B.F. Skinner)

This description of the writing process: "First you make a mess, then you clean it up." (W.G. Perry, Jr.)

"Write anyplace." (Natalie Goldberg, in Writing Down the Bones, a short and wonderful book, in answer to the question, "Where should I write?")

"When you're tying up your writing for the day, park on the downhill slope." (Ken Skier)

"Don't talk about your ideas before you write them down." (Joan Bolker)

Now to some more specific ideas for revising.

Here are some ways of moving from the messy zero draft form of your paper to a true first draft:

• Pick out anything in the mess that seems interesting, or strange, or provocative, or resonant to you, and try using it as a prompt to write deeper into your draft, i.e., literally put it at the top of the screen or page, and write, starting with a question like "Is there something here that will lead me forward?" Or "Does this contradict what I was thinking about these data?" Or "Curious — how does this bit reflect on my sense of this paper?"

• Ask yourself, "What stands out for me most in what I've written so far?" "Is there an argument in this mess?" "What point do I really want to make?" "Does what I've said here seem true to me?"

• Try doing short bits of freewriting, repeatedly, in answer to the question, "What am I really trying to say here?"

• All of the above questions and possible answers shouldn't go in the word balloons above your head, but instead, directly into writing; nor should you chat with someone about them before you've written them down.

• Probably the most important task, at this early point in the revision process, is learning to ask questions of your text, setting up a dialogue between you and the words you've written so far. At this stage it isn't useful to focus on getting it perfect: this is a time for opening up the possibility of new and divergent thoughts, not for narrowing them down. Right now you are the only audience for your work; later you'll have to think about your readers, but not yet.

• This may be the most important suggestion: Be sure to choose a working process that suits who you are, and not whom you'd like to be, or think you ought to be. Neat people and sloppy people, organized and chaotic, are each capable of producing terrific writing. As of this moment, there are no personality replacements available. In my other profession as psychotherapist, the work is to learn how to use who you are to have the most satisfying life you can; for you as a writer, the same holds true: come as who you are.

9. More About Critics, Expectations, and Ownership

I often have conversations with the writers I consult with about whom they're writing for. Recently one of them, an experienced doctor, asked me to read a short essay she'd written, and to tell her how to make it better. It was a nice piece of work, clearly argued, and interesting, but what struck me as I read it was that it was too polished. This may seem like an odd thing to say: don't we want our writing to go out into the world perfectly polished? Well, yes and no. This particular piece was about a dramatic change in the writer's perception of herself and the world, in the face of an illness that challenged her deepest notions of who she was. But while there were lots of interesting thoughts, there weren't many feelings in this essay about being rocked to her deepest foundations.

What I answered, in response to her question about how to make her essay better was, "Try writing this over, as if the audience were only yourself." Why this strange advice? Because what I thought I was seeing in this highly polished piece of work was what happens to a piece of writing when the writer focuses on her audience from the moment she begins it. In this particular case she began and ended with writing that, while about something quite personal, was impersonal: the author was there only as someone telling an interesting story, not as the main character in that story, with powerful and conflicting feelings; she put on her public face because from the very start she felt herself to be speaking publicly.

What has this to do with the sort of writing most of you are doing, much of it technical and data-based, and written on demand for granting agencies or professional journals? Or, to put it a mite more psychologically: your work may often be focused on others' (the patient's, the journal's, the institution's, the grantor's) needs, and not your own.

I want to raise the possibility that even with this kind of writing task it's possible to give yourself some room for owning your work. Why does that matter? Writing in which you allow yourself to have a personal stake (other than worries about what the reader is going to think of it, and you) more often, after revision, turns out to be both more enjoyable to do, and better writing. You may recognize some of these ideas as having been central to my argument for doing "messy writing." That's because writing for yourself is often less constrained, more divergent, and therefore, what you might call "messier."

But what I'm most concerned with here is what not claiming ownership of your writing, i.e., not writing for yourself, at least at the beginning of the task, does to your feelings about your writing, and your investment in and commitment to it. I think that it's probably possible, with most writing tasks, to find a way to claim them for your own, beginning by putting in everything you're thinking and feeling, no matter how outrageous, or contradictory, or maybe even wrong, just so you can explore what you might want to say. Since you can't get a personality transplant, you really have no choice but to come as who you are, at least at first. Once you've done that, there's plenty of latitude for polishing your writing, taking out the more outrageous witticisms or theories, or the nasty attack on another author that you've been wanting to make, and making the style conform to the acceptable form for a particular agency or journal or reader — because you've had the chance to write honestly, and openly, and expressively. Even after you've done all of the scouring of the text to make it presentable, you're much more likely to end up with a better and more lively piece. And I think you'll be more likely to return to writing with some sense of pleasure.

10. Holidays, Writing, and More About Critics and Ownership

About time and holidays: If you have time off, you need to use most of it for restoring yourself. If you feel you absolutely must work on a project over a vacation, I'd strongly suggest that you limit yourself to two hours a day, preferably first thing in the morning. Working this way has several advantages: you're much more likely to get to your writing, since you know it won't be interminable; the writing you do will benefit from your clearest thinking (most writers lose their focus after a couple of hours of intense writing); you won't resent the work for depriving you of much-needed time off; and you stand some chance of having some genuine rest.

But it's also fine not to write over a break. Restoring your energy and soul is very useful for being able to write later. Never taking any time off leads to an exhausted writer and tired writing. Here's a reasonable fallback position: just plan on writing for ten minutes a day, and no more, so you know you still remember how.

More about critics and ownership: I want to suggest something that I think could end up being very useful, not only to you. Let's go back to the dynamic of "writing for the critics" (posts 6&7), and narrow that audience to your department chair, or boss, or mentor. When I wrote my book for dissertation writers, perhaps the best suggestion I had from academic friends was to include a chapter for dissertation advisors. I'd like to be able to do the equivalent for you: to offer the people who assign, criticize, and accept or reject your writing projects some suggestions to help them be most effective and humane during this process, help you write for them with less anxiety, and help both of you get important work done.

Here are some related questions for you to consider:

• If your mentor is very supportive and encouraging, how would you describe how s/he does this? What, specifically, does s/he do or say that encourages you to write more and/or better?

• Do you have any trouble with the person to whom your writing goes, vis-à-vis criticism, or timely replies, or standards, or anything else concerning your work?

• What's most important to you in the responses you get from your reader? What sort of feedback has been most helpful?

• What is most disturbing/distressing to you in feedback you've gotten on your writing?

• In an ideal world, how would you like your writing to be received and responded to?

Finding some way to relay your answers to those who supervise your writing can make a great difference in the important and sensitive negotiations that go on between writers and critics. And if the critics are tamed, and the writers heard, I'm betting there will be more, better writing done, with less angst. Even the best-meaning, most sensitive reader and evaluator of your work cannot read your mind. Try to find ways to help that person help you more effectively.

11. What's The Goal of Your Writing?

Trying to understand the particular ways in which writing is stressful for many of you, I've focused more and more on the issue of time: given the workload you carry, which is of the "the faster I go, the behinder I get" variety, writing can't often be high up on your list of priorities, since it's not a matter of life or death (although perhaps a matter of professional life or death). It's probably only when a piece of writing is due, or overdue, that it gets to occupy a place higher up on your must-do list. So, if someone like me suggests that absent an emergency deadline, you might want to think about the writing process, and perhaps even to spend some of the time you haven't got learning a different, more productive way to do it, that can't be a terribly appealing idea. This "no time" diagnosis was my first.

But recently I've been thinking about another possibility, one hidden behind the "no time" explanation. I've been trying to teach you two principles, the first, that writing is a matter of "first you make a mess, and then you clean it up," the second, "write in order to think: you don't have to have everything thought out in advance of writing." These two principles are clearly connected: if you permit yourself to "write in order to think," it's likely to be a messy process at first, one that requires several iterations to get to a finished product. (My son teaches mathematical biology and epidemiology, and is trying to help his graduate students improve their writing as well. When I talk with him about the make a mess/clean it up process he tells me, "I'm sometimes worried that people don't recognize very well what the endpoint is." You have the opposite problem, of hurrying too quickly to that place.)

Which leads to my next thought about the disease of "no time." Even though you've been telling it to me, in one way or another, all along, what I now hear is how often — nearly always, in fact — when you describe the projects you're working on, the focus is on deadlines, on "getting it done." Far be it for me to discourage writers from getting it done! But I worry that when this is the primary goal of a writing task, both the piece of writing and the writer suffer. Writing only to

39

get it done short-circuits the thinking that makes for a deep and rich piece of work, and certainly does away with any enthusiasm for the job.

I hope that at least some of the time you're writing about questions that interest you, and not just because the task has been assigned, or you've promised it to an editor, or your promotion depends on it. If the questions you're addressing are important to you, and it matters to you to be moving towards answering them, and feeling that you're onto something, focusing on the endpoint isn't the most productive way to get there, at least not until you've first done a lot of exploratory writing. If you can bear it, what is much more likely to get you the result you want is to move your attention from the endpoint, to the process of searching out what you might know, and can say.

Try this experiment: Pick the piece of writing on your list of work to be done that most interests you for its own sake, and try putting any deadline for it away for a while (clearly this won't be the paper you need for tenure!). Begin working on it by doing as much daily messy freewriting about it as possible, given your schedule (even if you only have ten minutes). Ideally, do this writing before you tuck into the rest of your day's work. Try this for a week or two.

If you've found ideas you want to develop popping up in the freewriting, or questions you want to pursue, switch to iterative, or directed, freewriting, writing as quickly and uninterruptedly as you can for at least ten minutes. Put the questions or wacky ideas at the top of the paper or the screen, so that you get down on paper, or on the computer, any neat or messy, confirming or contradictory, mainstream or way-out thoughts that are related to the ideas that emerged in your original freewriting. When you've accumulated a significant amount of mess, call it a zero draft, and then prepare to pick out the usable pieces for the drafts that will follow.

This method is clearly not for case reports, or the results section of a paper. But it can be immensely useful when you come to the part of your paper that involves making sense of those data, i.e., the interpretation part, the piece that contains the interesting questions and then answers them.

Try this, and see what happens.

12. What's the Diagnosis, What's the Cure?

Sometimes, as I work at understanding the writing culture at M.D. Anderson, I feel a bit like a creature from outer space. I think I may be getting closer to your world, so today I'll gather my courage and offer a few observations and a tentative diagnosis of the problems I've been seeing.

In the last post I wrote about the problems created by what I see as the propensity to focus on "getting the writing done," rather than on doing it. This focus often, paradoxically, ends up with your getting less done, with the exception of worrying — you do a lot of this. The pressure to get writing done exists for many good reasons: you have important research results to make public as soon as possible; you have many simultaneous tasks to do, and getting items on your list checked off reduces the burden on you; you're simultaneously working in a medical and an academic institution, where "publish or perish" (or, at least, not get promoted) is a given.

And some of you have had narrowly scientific educations, in which writing wasn't taught, and was encouraged only in an endpoint sort of way, as a tool for getting papers out.

I suspect, too, that you may see writing for its own sake as a frivolous activity, one that can hardly measure up to the work you do that concerns life and death. Maybe you think that it's "art," while your other work is "science." But I'm not proposing writing for the sake of writing, even though it may look that way.

Given all of these forces at work, learning a new writing process may seem like just another item on your list, maybe an adventure that would interest you if you had a different life, but right now something that feels much less important than the other work you're trying to do.

I'll admit that I have a prejudice about writing (as well as a lot of experience teaching many different sorts of people how to do it), but

I believe that if you will venture to devote a *very* small amount of time to this project, the return on that investment will be large. My aim isn't to try to turn you into poets, or novelists, or even bestselling medical writers (if you're already a member of one of these groups you probably know some of the things I'm trying to teach). It's to help you develop an individual writing process that will allow you to write faster and more easily, with better results, to get your research out into the world not only more quickly, but also in a form that's much more likely to be read.

Which brings me back to the question of freewriting. How can something that looks like such a mess, that has so few rules, be the beginning of a process that will yield fine, clear writing? Sometimes I think the transition from freewriting to iterative writing that's revised (literally, "seen again") over and over must seem, if you've never experienced it, like an impossible leap. I need to explain it in a way that makes it clearer, and not a magic trick.

This transition is about the move from play to getting down to work, from free association to the attempt to begin an argument, from divergent thinking to focused thinking, from an absence of rules to beginning to impose structure, logic, and plot on the mess of ideas that emerged in that space of freewriting. Another way of describing the two different processes: freewriting engages the creative mind; iterative revision engages the critical mind. Both play an essential part in producing excellent writing.

But in order to engage with all these complementary processes, you need to begin by committing yourself to a very small amount of time in which to do freewriting, as many days of the week as you can possibly manage. You'll very likely discover that this will free up your thinking, and produce some good ideas you might not have gotten to if you relied on the usual "I've got to begin by getting this piece done" method. Yes, freewriting produces a lot of junk, but think of it as compost — or as fertilizer, if you prefer.

Next time I'll write about harvesting and cultivating the ideas that emerge, and using them to produce something you can honestly call a draft.

13. How Good Is Good Enough?

I had a conversation recently with someone who'd written a short scientific paper describing a study he'd worked on; his concern was that it wasn't "good enough." As I read the paper I noted it was clear, succinct, and well-organized, presenting data in a way that made it very easy to understand. I asked what he meant by "not good enough," and he talked about how maybe the style ought to be better.

One of the pleasures of my work is learning a lot from the people I teach. This writer's concerns taught me that I needed to say more about the purpose of and the audience for the writing you do. What I said to him was, "This paper seems to me to do exactly what it needs to do, it's good enough, and it doesn't need to be held to the writing standard of *The New Yorker* magazine."

There are at least three lessons to be taken from this interchange: first, that you have to consider the goal of each of your writing tasks; second, that you need to pay attention to your audience; and third, that you have to think about how much of your energy and your heart a particular piece of writing is really worth to you.

I know that you often have several writing tasks at once on your to-do list. One of the risks of this is that you'll give all of them equal weight. But some tasks are important, some are trivial, and most are in between. If you have a grant proposal to write, one that could produce support for your research for a major chunk of time, it's very important that it be right, and compelling, that it say to the grantor, "You'll regret it if you don't fund this amazing project." But if it's in the right format, and you make your argument well, it doesn't have to be at the stylistic level of writing by Shakespeare or Jerome Groopman or Oliver Sacks. If it's clear, it's probably already better than many of the proposals it's competing against.

If you're writing a paper that reports data or results, your reader has to be able to read it with relative ease; the tables have to be constructed well, and you have to remember that what you're writing about isn't necessarily obvious to the person reading it (if it were, you wouldn't need to write the paper). (George Gopen and Judith Swan's article "The Science of Scientific Writing" has some very interesting things to say about writing for the audience.) But even though you need to be careful to consider such things, you don't need to write a perfect paper (there's no such thing).

I don't mean to suggest that standards don't matter, or that you should send out a piece of writing that's incomplete, or sloppy. What I'm arguing is that not every piece of writing weighs the same, or is worth the same level of commitment on your part.

How do you decide how much effort a piece of writing deserves? Clearly, you don't spend as much effort on an email as you do on a paper you're going to publish (although I've learned from experience that it's useful to reread my emails before I press "send"). And you don't, most likely, need to use as much energy on a paper that you'll give at a conference (unless it's reporting groundbreaking results) as you will on one that will be published in an important journal. If you're writing a book, or part of a book — something that you want to be able to see in print without wincing, even if you reread it years later, writing that you hope will have permanence in the literature of your field — you still won't be able to make it perfect, but you'll probably want to make it as good as you possibly can (and this will most likely involve having trusted readers/editors read it over before you declare it finished).

Remember, "life is short, and the art long," to quote Hippocrates, who's quoting Seneca, and you need to apportion your time and energy in proportion to the importance of your task.

14. Writing In Good Company

Writing is a lonely sport. Some people enjoy this aspect of it, and others find it very dispiriting, enough to make them avoid writing altogether. You may be one sort of writer at one stage, and another at a different part of the writing process: you might like to draft your first thoughts away from the world, but hang out in a library, or a coffee bar, or with a friend, when you're revising, or vice versa.

For better and worse, though, writing is never a totally solitary sport. Even if you're closeted, away from email, phone, and people coming to ask you questions, you're generally "with someone": the ultimate audience for your writing, or in memory, with the 7th grade teacher who told you you could or couldn't write well, or with your own internal critic or judge. I suspect there are very few writers who would bother to write if they never hoped, or expected, their writing to be read by others. What do you make of all the company that accompanies your writing attempts?

What matters most is that you write in good company. Write for those whom you know will understand you, and be on your side. They will offer useful criticism if it's called for, but not overwhelm you with it, is as good as it gets. It may still feel anxiety-provoking, or embarrassing, to hand over a piece of your writing. How much it does will depend on how finished your piece is, and/or how anxious you are. But writing by its very nature exposes the writer, even if s/he hides behind the third person.

The writing project that's going to be read by someone whom you know is mean-spirited or hypercritical is harder. (I think that a rejection by an unknown journal editor who turns down your article is much easier not to take personally.) What might you do in this situation? It helps to find another, trustworthy reader to show your work to first, or simultaneously; if you can avoid it, don't hand an incomplete or unrevised piece to the nasty reader. It can help to talk to yourself about how hard it is, but also to follow the "writing well is the best revenge" strategy.

What you can't afford is to allow the difficult reader to take over your writing and your own opinion of it. (If *all* of your readers feel this way to you, you need to take a good long look at yourself, in one therapeutic way or another.) If you have the sense that the difficult reader is clueless about his/her effect on you, rather than mean, you might try summoning up your courage to say that it's a bit hard for you to listen to his/her feedback, but that you'd find it easier if s/he could send the comments by email, so you could digest them a bit more slowly, or avoid writing on your ms. in red ink, or....

The worst possible companions, though, are often ourselves. Imagination is wonderful, unless we turn it to spinning scenarios where people say godawful things about our writing, ignore it, don't fund our proposals, or tell us our data don't show what we've said they do, etc. (I'm sure you can easily add to this list).

Years ago, with one of my first clients, I realized how powerful the imaginative self-critic can be. He came to see me because he couldn't write at all, and desperately wanted to. I asked him some questions, including what sort of setting he wrote in, and he mentioned that he had a painting of one of his illustrious ancestors hanging where he could see it while he wrote. This ancestor, who resembled one of the vicious characters in the paintings at Harry Potter's school, seemed to be spewing invectives, telling him what a worthless character he was, and who was he, anyway, to think he could write anything worth reading? The fix was apparent and quick: we decided together that he'd turn the ancestor's picture to the wall (this somehow seemed more appropriate than removing it altogether). My client became a quite prolific writer.

Some of us figure we're better off criticizing ourselves before someone else does. Some of us were taught that the critical mode should always be the most important one when we write. But if it's applied too early, it can destroy thinking and creativity. Some of us are frightened by the extent to which even "objective writing" (if

50

there is such a thing) exposes us, so we scare ourselves away from the task.

Several of the writers I've worked with got over writing blocks or writing procrastination by finding writing partners or groups to work with. One of my clients, who choked each time she faced a deadline, had a good friend who was a professional writer who invited her to come to her house and sequester herself to write, knowing that this friend was in the other room cheering her on — and writing.

One of the many good things about writing partnerships is that you choose your company. Another is that the partner isn't there to judge your writing, just to egg you on. Or, more accurately, you egg each other on. Getting out of solitary confinement can also turn down the nutty negative voices that often emerge full-blown when you're alone. You have a simple obligation to a writing partner: to show up, and to write, and only that.

It's hard enough writing even if you love it (some people do), even if you have a benevolent audience waiting to see what you produce. It's very useful to work hard at finding the best possible company — both in yourself, and in others — to support this rich, complex, and difficult enterprise.

15. Writing Travels

When you have a writing routine at home, how do you translate it to a workable method for writing while you're traveling? Many of you travel often for work, but I'm going to begin by talking about the sort of interruption that vacations bring to whatever routine you've established.

First thoughts, maybe obvious, maybe not: everyone (even you) needs, and is entitled to, some time off. Those of you who grew up outside the U.S. know that Americans have much less vacation time than, say, Europeans. If you've planned on taking two weeks to spend with family or friends or yourself, don't expect to get any writing done; don't even bring it with you. (If you feel that you must, don't plan on writing more than half an hour on any given day; even better, write the story, or poem, or letter you've been trying to get to.)

Set a deadline before you go away. Writing is something you don't want to worry about while you're on vacation, trying to relax and restore yourself. At the least, "park on the downhill slope" before you leave: write out for yourself where you've gotten to on your project, and what you hope to do after you return. If you have interesting ideas or hunches you'd like to explore, jot them all down. This way, when you come back to work you'll know where to start back into your writing.

But I suspect that most of your travels aren't for vacations. I know from our correspondence that your professional trips often interrupt projects you're working on. Here are some suggestions for getting writing to travel:

• Lower your expectations of how much you can accomplish, or how much time you can spend writing while you're away. Writing for half an hour a day is much, much more useful than promising yourself you'll do a few hours each day, and then not getting to it — because it's very hard to find that much opportunity to write while you're traveling.

• Writing first becomes especially important when you're traveling for work. The surest way to find that half hour is by doing it before the meetings, conference sessions, or other obligations eat up your day. Write before you leave your hotel room, or hide out in a coffee shop where you won't meet any of your colleagues.

• Be flexible in how you write: besides your laptop, bring a pad or notebook (and a non-leaky pen, or a pencil). (I once had a student who used the brown paper towels available on airplanes for writing while she traveled, because "I didn't feel like I was wasting good paper on my messy writing.")

• If you miss home, see if you can use your writing as a way of feeling more at home, as a respite from whatever busyness or business you're involved with on your trip.

• Traveling might actually be a good writing opportunity *because* it's different from your usual writing routine.*

• If there's no good place to write where you're staying, find a coffee- or teahouse where you can sit anonymously and write for half an hour. (I expect that if I were to travel to Antarctica, I'd be very likely to find a Starbucks there...)

• Redefine the purpose of traveling writing: think of it not as a grand opportunity for finishing an important paper (unless you're one of the rare souls who's actually been able to do this), but as a way of keeping your project simmering, so that you can return to it easily when you're back home.

• Insofar as it's under your control, don't schedule major travel when you're up against a writing deadline.

• If you can't use your traveling time to do focused, goal-oriented writing, use it to back off and get some perspective on where you are with your projects.*

• Sometimes, if you're lucky, traveling can give you *more* space, if you can find a way to use it. Being away from your normal responsibilities can turn out to be a boon for your writing. (If this is so, ignore the half-hour limit above!)*

• If writing with company is something you enjoy and find productive, try to engage a colleague you're with during your trip in writing with you, and so increase the chances you'll both get work done.

* Thanks to Ben Bolker, my traveling scientist son, for these additions.

16. Making Space for Writing

Consider first the actual place where you write. How do you decide where it will be? There's no "one room fits all" answer to that question, but an anecdote will give you an important clue. Years ago, a doctoral student who consulted me because he couldn't write his dissertation answered the question "Where do you do your writing?" by saying, "I always write at the kitchen table." I said "That's fine — how much do you get done there?" He answered, "I never get anything done there." I then suggested he find a different place to write. If you know what your equivalent of the kitchen table is, do make sure you go someplace else to do your writing. (And, not to hold the poor guy up as a laughingstock, I'd be willing to bet that many of us often unconsciously choose our own version of the kitchen table to write at, and if this is what we're doing, we need to notice, and move.)

So, where might it be good to do your writing, and how do you figure this out? Some people like anonymous, crowded places to write in (some of them even like them noisy), others would rather sequester themselves in quiet, private rooms. But there are some elements that seem to matter for everyone.

Think about where you've done your best, most productive writing, and why that was. Did it have to do with the place, or the company, or just with the project? Are there some places where you rarely get good writing work done? (Anyplace where you're frequently interrupted by people, email, or the phone probably fits in this category.) Do you need quiet in order to focus? Does a good set of headphones, either of the noise-canceling sort, or ones attached to the music that best keeps distractions at bay, help you write?

An important subset of the "interrupted by other people" scenario: If you're trying to write with young children around, wait until they're reliably asleep, or get up before they do in the morning (I can hear someone saying, "Right, at 5 a.m.!"), or find yourself someplace outside your home to work. Trying to write around kids who need you isn't fair to them or to you.

Do you prefer windows and light, or are you better off in a room without a view? Do you like to be surrounded by other people who are involved in writing (in a university library, for example), or to have a writing buddy with whom you share a vow of silence while you both work? Figure out what you need: anonymous noise or absolute quiet, a public place or a private one designed to suit all of your personal preferences. You may already know the answer to this question, but if you don't, try experimenting, keeping careful track of the results of, say, working in the library versus in a private place.

And then find or create such a space, and write there. I've known someone who borrowed a spare room in a friend's house, another whose colleague loaned her his library carrel, someone else who created a soundproof study in his basement. The hardest part, I suspect, is figuring out which sort of space will be best for you to write in. (And your preference might change over time: check this periodically. When I was an undergraduate, I wrote most of my course papers in bed; now the very thought of it makes my back ache!)

Figure out a prompt that helps you psychologically to enter the space you've chosen: soothing music, or five minutes of yoga, or Tai Chi, or having left your desk clear so there's no clutter to plow through to get to its surface (the last of these is only for people who are bothered by clutter). In the same category, remember to "park on the downhill slope," so you know where you left off in your writing, and can more easily get back in.

Writing space is linked to time, of course, as in, "When, in my very busy life, can I find the real and psychological space that I need in order to do my writing?"

I can imagine you asking, "But what about on call, or stat messages?" Obviously you must tend to them — and this is a strong argument for seeing if you can write before beginning your official work time, and for knowing, as well, that you're unlikely to get much done during on call time.

17. Looking Again At Your Goals

The goals for this book are pretty clear: first, to help you produce more writing; second, to do that more quickly; third, to do it with greater ease; fourth, to improve the quality of your writing; and finally, fifth (perhaps the most important of all), to help you use writing to expand your thinking.

1. How do you produce more writing?

— By figuring out how to use time you hadn't realized was available for writing (e.g. early morning, or ten minute snatches); by paying attention to the circumstances in which you write (so, not "at the kitchen table where you never get anything done," and not in places where people feel free to interrupt you); and, paradoxically, by lowering your standards for what constitutes "real writing."

— By choosing the mode — handwriting or writing on a computer — that helps you avoid interrupting yourself to make corrections or obsess about word choice as you write early drafts.

2. How do you write faster, and 3, with greater ease?

— Once again, by lowering your standards — but not really, because fast writing in a first draft is often at least as good as writing in which you've paid attention to every single word and punctuation mark. (Here's an easy trick for fast writing: if you can't find the exact word you want, put the nearest one that comes immediately to mind in brackets in the text, and then figure out later, while you're revising, what the best word is for what you're trying to say.)

— Write faster by treating it as a race with yourself: "How fast can I turn out two messy pages?" I learned to write course papers quickly in college by setting myself the goal of writing five pages each day. I'd revise them later, so there was no sense in "putting in time" on an early draft, and if I could write those five pages quickly, I freed up time to do other things – including later revision.

— Write faster by learning to not let yourself stop writing, even if it's junk writing. When I began a writing center many years ago, a colleague and I agreed that the most important tool for a writer was a pot of glue: first you put it on your chair, then you sit down.

You may be wondering how the goals above can lead you to produce better writing; here we get into deeper water. I've worked with many blocked writers, and when I've asked them to experiment with writing quickly and nonstop, I've seen them produce high quality writing that surprised both them and me. People who learn how to write quickly (with, of course, the caveat that such writing is the first part of a two part process, and will require revision) also often turn out to write better. Fluency helps quality.

This takes us to the most grandiose goal of all (but I know it's attainable): learning to write in order to think. This idea may violate everything you were taught about scientific writing ("first get your data, then write it up"). Writing in order to think isn't some sort of magic. It depends on the fact that we often know more than we think we do, because not all of what we know is stored in the most accessible part of the brain. Thought depends on imagination (the biographies of great scientists demonstrate this over and over), and imagination isn't a matter of plodding, but rather, of freeing up the mind to wander and play, to give it the opportunity to see things differently, and so, to make new connections.

18. Rotten Rejections

One of the many hazards of professional writing is the disheartening, and sometimes harsh, rejection of an article or a grant proposal. I want to consider some of the possible ways of handling such rejections, and how you might understand their meaning.

First, a disclosure: nearly every manuscript I've ever submitted was rejected before it was published, some of them multiple times. The first time this happened was with a short essay. I was devastated by the rejection letter, in which the editor not only spoke poorly of my essay, but also impugned my character and intelligence. Fortunately, I sent the essay to a second (and more prestigious) journal, whose editor accepted it quickly, and praised it highly. (I still get a sweet sense of revenge each time it's quoted.) My dissertation-writing book, a bestseller in its market, was initially turned down by nearly a dozen publishers. A few such experiences with publishing have taught me some useful ways to deal with rejection, which I'll pass on to you:

1. A reviewer is just one person, with quirks and preferences, fondness or dislikes for particular substance and styles, some of which may be idiosyncratic. (And there's always the chance that s/he had a toothache the day s/he read your paper.)

2. One person's opinion is just that, and no more.

3. Far more papers and proposals, no matter how good, are turned down than accepted. But gratuitous nastiness on the reviewer's part is always inappropriate.

4. Papers that put forth new ideas, challenge old ones, or are provocative in some way, often elicit negative or conflicting reviews.

5. Submissions that are turned back with suggestions for revision haven't been rejected. Swallow your pride, see if the suggestions work, and keep your mind open to making changes that will improve both

your paper and its chances of acceptance. (Think of the reviewer's comments as doing half the work of another revision for you.)

A few suggestions:

• Don't expect your paper or proposal to be accepted the first time you send it out.

• As a hedge against the strong possibility of a rejection, before you submit your work the first time make a list of several journals or agencies that might be good places for it.

• Then, if your paper/proposal is turned down, pick yourself up off the floor, see if any of the reviewers' suggestions need to be incorporated, and if so, do that, and resubmit the work a.s.a.p.: don't put it away in a drawer because someone didn't like it. Sulking, or beating your breast, isn't useful at such times.

• Be realistic about the odds for grants being funded, and try not to take the rejection of a proposal personally. Keep trying. (One of my husband's colleagues, Robert Morris, offers this mantra: "If you're getting most of the grants you've applied for, you're not applying for enough grants.")

When I first began submitting my own writing to journals, a poet friend told me her strategy: "I make a list of ten places I might send my poem, and as soon as it's been rejected by the one at the top, I look at it to see if I want to make any changes, then put it in an envelope and send it out to the next name on the list — and add another possible journal to the bottom of the list."

It's hard to take rejections well, but practicing doing so, and assuming that the turndown is not the end of the world, nor the last possibility (a.k.a. the "get back in the saddle" strategy), will help you be less devastated.

The title of this post comes from a book called <u>Rotten Rejections, A Literary Companion</u>. It's a book I've often used for consolation, a sampling of classic rejections. Here are a few of its gems:

re: Joseph Heller's <u>Catch-22</u>: "He has two devices, which he works constantly...This, as you may imagine, constitutes a continual and unmitigated bore."

re: Nabokov's <u>Lolita</u>: "I recommend that it be buried under a stone for a thousand years."

re: Pearl Buck's <u>The Good Earth</u>: "(We) regret the American public is not interested in anything on China."

Maybe my favorite rejection in this anthology, from a Chinese economic journal:

"We have read your manuscript with boundless delight. If we were to publish your paper, it would be impossible for us to publish any work of lower standard. And as it is unthinkable that in the next thousand years we shall see its equal, we are, to our regret, compelled to return your divine composition, and to beg you a thousand times to overlook our short sight and timidity."

And you thought you had it bad....

And one more: Madeline L'Engle's <u>A Wrinkle in Time</u>, one of the most popular children's (and adult) books ever written, "was rejected by virtually every major publisher" (26 of them!) before being accepted by Farrar, Straus & Giroux. It's been in print for more than 50 years.

19. Reading in order to write

Here are some thoughts on the writing/reading connection, as well as a list of a few books that might be useful:

One of the things I hear often from the writers I work with is that they need to read more before they start writing. To rewrite a line from the Book of Ecclesiastes, "Of the reading of books (and of papers) there is no end." How do you read enough of the writing on your subject so you know what has and hasn't already been written about it, without devoting your life to chasing down every word on the subject?

Let me switch for a moment to my psychologist role, and explore why people seem so obsessed with reading more — and then yet more — before they begin their own writing. I offer these possibilities:

• It's easier to read other people's stuff, and feel like you're working on your project, than it is to begin putting your own words onto a page (or a screen);

• You have a terror of publishing something that's already been done, and fear that if you don't read everything, you might miss something that proves your work wrong;

• You have the (mistaken) belief that everyone else out there is smarter and knows more than you;

• You may have a mistaken notion of what it means to survey the literature.

Years ago I used to wander through the stacks of one of the largest university libraries in America to find the shelves of books on the subject I was about to write on, and felt I needed to read every single one of them. But as a graduate student I took a very boring English course that provided me with one experience that made sitting through it worthwhile. Each week the professor would look around the room, point to someone, and say, "By next week you will present an annotated bibliography of all the literary criticism about x's work." When my turn came, it was to find all the commentary written about Ralph Waldo Emerson's poetry. I was terrified, but because the time was so short I had to get to work immediately, and learn how to read and assess and summarize all of the relevant papers and books.

I discovered that one could figure out quite quickly what was worth including in such a bibliography by skimming the title, the introduction, the table of contents, and, in a book, the first and last chapters and index. In a paper, the abstract and the first and last paragraph were sufficient to tell me whether I needed to read further. And what's more, I discovered that not all of the things ever written about Emerson were worth citing.

Knowing how to read to evaluate a piece of work quickly is a highly useful skill, and one that isn't too hard to learn (although I suspect it goes against the very careful reading that scientists do much of the time). And if your quest is really for the papers that you must read (rather than an excuse for not doing your own writing), you'll find them relatively quickly.

Let your writing guide your reading, not vice versa. Here's another way of looking at the "how much reading is necessary?" question. If you set out to read everything ever written about your topic, you'll have an enormous job ahead of you. If, instead, you begin by writing down your own questions and hunches, and what you already know, you'll be able to narrow your search and make it much more efficient. Your own questions will lead you to the literature you really need.

Why reading is useful:

I'm not about to unsay what I said above. I want to look briefly at another kind of reading, the sort that's not strictly relevant to the topics you're writing about.

The main question I'll try to answer is, "Why spend any of my rare and precious free time reading stuff that won't advance my work?" The answer is, "Because it can help and improve your own writing."

Reading the work of a talented writer may resemble listening to a talented musician: if you hear enough good music, you come to know what "good" means in that realm. And if you read good writers, of science but also of anything else, you'll come to "hear" the difference between passable writing and really fine writing. (If English isn't your first language, this sort of reading is also a good way to improve your skills in both speaking and writing.)

You may also find that the standards for style are more open that you thought (I remember when *Science* began permitting the use of "I" in their pages). Some journals include personal writing, as in the "Perspectives" section in the *New England Journal of Medicine*, and "On Being a Doctor" and "On Being a Patient" in the *Annals of Internal Medicine*. I'm a firm believer that the more you write anything, the better the rest of your writing becomes. (I once had a superb anesthesiologist who calmed me during a two-hour surgery by talking about the haiku he was learning to write — yet another use for writing!) And the more you read a wide variety of other people's writing, the more relaxed and proficient your own is likely to become.

Here are some books you might want to explore, by a bunch of very talented writers, some of them doctors and scientists, some not:

> Iain Bamforth, M.D., editor, <u>The Body in the Library: A Literary Anthology of Modern Medicine</u>
> Tracy Kidder, <u>Mountains Beyond Mountains</u>
> Michael A. LaCombe, M.D., editor, <u>On Being A Doctor: Voices of Physicians and Patients</u>
> John McPhee, <u>Heirs of General Practice</u>
> Lewis Thomas, M.D., <u>The Fragile Species</u> (or anything else by him)
> Berton Roueche, <u>The Medical Detectives</u>

Articles by Jerome Groopman; anything by Oliver Sacks; *The New Yorker* magazine (often there are terrific scientific or medical essays, and if not, the cartoons are spectacular); *The New York Times* science pages.

The moral of the story is, read: to find out what's permissible, for pleasure, for improving your language skills, for relaxation, for learning more about writing, and for enjoying the sound of the words.

20. Writing, Interrupted

This post focuses on interruptions to writing, and how to get restarted. Interruptions come in several different forms. I'll begin with the simplest variety, and then move on to some more interesting species.

If you find it hard to re-immerse yourself in your writing projects after some time off, here are some possibly helpful hints:

First, can the guilt. It's not reasonable to expect that you'll work all the time while ordinary people are partying. More important, though, guilt is a kind of negative reinforcement, and so, not at all helpful for reestablishing your writing habit.

Second, start back to writing gradually. If you give up exercising, you have to begin again slowly, because you're out of condition. You need to do the same with writing, beginning with short, frequent writing sessions (daily if you can manage it), and waiting for the spark to catch — which will happen.

Third, pick the easiest writing task on your list to start with, not the hardest, nor the one you most dislike. If you do the latter, you're choosing self-punishment, and it will work to "extinguish the behavior," as negative reinforcement often does — in plain English, it will stop you from writing.

Now we get to some deeper sorts of interruptions to writing, what I call creating interruptions, or inviting them in.

How do we try to escape from the hard, and at times anxiety-provoking, aspects of writing? There are many truly brilliant ways — in the last post I talked about reading as a way to avoid writing, consoling oneself with having read everything ever written about your subject. Spending too much time on outlining, or on struggles with coauthors, are some other activities that don't produce actual writing, but convince you (sort of) that you're really working hard on your task. These are among the many possible ways of interrupting writing.

During my years of working with writers I've seen lots of ingenious methods, but perhaps the most dangerous is "working hard at writing" by just going through the forms: focusing on organization, or tables, or other parts of structure, and not attempting the much more arduous work of using writing to think. This kind of self-interruption can wreak havoc — and is often not totally conscious.

Another mode: inviting chaos into your life. (This is not the same as having a busy life.) Deciding to move houses, or to renovate an existing one, as a writing deadline draws near, is a bad idea (assuming you have some choice about the timing). Feeling you must begin a new research project when it could be realistically postponed until you finish the six projects already on your desk, or accepting every new work opportunity that presents itself, are complications you can regain some control over. Here's a sneaky way to do this: Promise yourself that when someone asks you to do something you may say "no" immediately; but you may not say "yes" without telling that person "I have to think about it, and I'll let you know."

Finally, one of the most prevalent interruptions to writing is what I call "mind static": the lists, preoccupations, anxiety, or obsessions that draw your attention away from the work at hand. I don't know anyone who has a simple life, and there are always distractions, ranging from the simple "what shall I get for dinner tonight on my way home?" to "my coauthor isn't holding up her end of the work," to "I'm not sure my data have shown what I expected them to — and besides, what prompted me to choose this line of work?" This static can break an important train of thought, or frazzle you so much that you can't sit at your desk. I won't attempt to diagnose its causes, but want to offer instead some possibly useful ways of dealing with it.

First of all, it's important to know that our minds don't work in a straight track: thought processes are closer to the sort of narrative that James Joyce writes than they are to formal scientific arguments. A neat linear argument is the result of many drafts, some of them carried out in your mind, and others on screen or paper.

Is it better to try to put the static aside, or to pay attention to it? That depends on what it is. The third case above, for example, of worry about whether your data support your hypothesis, probably deserves attention (unless you recognize it as a neurotic tic that often surfaces in your work).

And there's a central aspect of static that I described in my book about dissertation writing:

"... some of the static is about a much deeper part of being a writer: there is something inherently and wonderfully selfish about claiming time for one's own thoughts and words, for taking them seriously enough to dedicate a ... piece of your life to them, and ... a smaller piece to the needs of others."

(Clearly, if you really dislike writing, the above doesn't apply to you!)

You might want to try focusing your mind first, and seeing if it helps (writing quickly encourages this). Another approach borrows from a Buddhist meditation practice, "training the mind-puppy not to wander off." If you've had any experience with meditation, this might be something to try with your writing.

Here are some ways of giving static its due without allowing it to disrupt your writing: have a paper and pen (or a separate computer file, if you insist) at hand, and write down your worry, briefly. Or have a notation (for me it's been square brackets) that can go into your draft, and indicate that what's inside isn't part of the work. (It's also then quite easy to delete later — just be sure you do!)

Write very quickly, and write first, promising yourself (perhaps by putting the issue your static presents on a pad nearby) that you'll deal with it just as soon as you've written some number of lines.

For all of the various self-interruptions that writers use, these strategies can work:

Ask yourself if the work you're doing is actually advancing your writing, or an escape from it, and then act accordingly.

Interrupt your work purposefully: take a walk, get a snack (forget alcohol: despite all the romantic stories about drunk famous writers, it's never been known to improve either the quantity or the quality of writing), and then come back and set a reasonable short-term goal ("before I leave today I'll have at least written x...").

Don't be overly impressed by the static: fight back with the strategies that work for you.

21. Email: Gift or Curse?

I hear a lot from writers about email, ranging from "[email is] very beneficial to me...it allows people to be available on their own schedule," to "the most inhuman way to communicate — I wish it did not exist," from "SPAM is the 'four letter word' of the new generation — how can filters keep out people you want to hear from and not the ones whose names are all consonants?" to a summary of David Allen's seminar about productivity (his book Getting Things Done has a few good pages about how to sort and file email and keep your inbox from overflowing — though it isn't for the technologically challenged).

So what do we do with this mode of communication? Listening to some of your feelings about it, I wonder what people had to say when the telephone first allowed other people (not yet telemarketers) to intrude on our privacy and quiet. So here's my first point: remember that email has replaced the telephone as a major means of communication, and that it permits you to read and answer it at the time of your choosing — if you can think carefully about when you want and need that to be.

Suggestion #1: Consider setting up specific times during the day when you read your email, and then turn it off. Don't make it too easy to check it dozens of times a day, because this is a real time-waster. People who need to reach you in an emergency will text or phone.

This leads me to a relevant psychological observation: Years ago, B.F. Skinner described intermittent reinforcement as the most potent kind. Surprisingly, rewarding an action sometimes, but not every time, is a better way to make that action more frequent than rewarding it consistently. If you think about email in this way, you realize that it's a perfect example: while your 20 new messages may be mostly spam, you might also get something important from a colleague whose work you've been waiting to see, a heartening letter from a friend or an offspring, or the notice of an important event — the last three being "rewards." What this means is that it's very, very easy to become addicted to email, even while you're cursing how much of your time it takes up.

I've worked with people who had busy lives, and good, interesting work to do, who spent inordinate amounts of time playing solitaire on their computers. My advice was obvious: remove it. But you can't remove your email. You can, though, set up some rules for yourself, in addition to the one suggested above.

• If you really can't resist email's lure while on your computer, try writing by hand.

• Think about email as a possible addiction, and consider what you've done with other things in your life that were in that category to get them back under your control — then translate your solutions to this addiction.

• Ask yourself if even while you're ranting about it, you're inviting email to interrupt you, because it's an easy and ever-present way of putting off other, more difficult work (particularly writing).

• Know that you own your time (or at least some of it), and don't assume that just because someone writes to you, you've got to answer on the spot.

• Get into the habit of filing, or forwarding to yourself, mail that you will want to respond to (just not now), as a way of not getting waylaid by items that don't really need your immediate attention.

• Use another one of the insights of behavioral psychology, and tell yourself you can't read your mail until you've done x (something you really must do): that is, turn email into a reward.

Think hard about how much email is really the villain, and how much you enjoy it, or enjoy its interruptions, and take action accordingly. Remember how magical email can be. It allows us access to colleagues and ideas we would have had to work much harder to get to without it. It allows ongoing conversations with friends, family and colleagues at times of our own choosing. And if you think hard about how you might use it, rather than allowing it to take over your time and life, you can harness its power for the best reasons.

P.S. For a delightful essay about email, see Anne Eisenberg's "E-mail and the new epistolary age."

22. Writing What You Want to Write

There are several possible meanings of "writing what you want to write." Perhaps the most extreme meaning is "I don't want to write any more papers or grant proposals; I'd like to write fiction." But between that extreme and its opposite, "I'm writing exactly what I want to write in my papers and proposals," there's a lot of territory. Let me describe some of it:

• You wish to write different sorts of scientific papers and grant proposals, devoted to the questions that interest you most, even though they may be less likely to be published or funded;

• You'd like to write in a different style, one that's more relaxed, or less stuffy;

• You'd like to write single author papers, to have the chance to create them without having to contend with the quirks or opposing opinions of coauthors, or the organizational quagmires of joint work;

• Or, you'd like to write more joint papers, to make writing less of a solitary occupation, to expand your knowledge, or to balance the weaknesses and strengths of your writing with the different ones of your coauthor;

• You want to take a shot at a really wild paper in your field, to see if you can make it convincing and publishable;

• You'd like to write about medicine/science, perhaps in essays like the ones that Jerome Groopman, Herbert Benson, and Lewis Thomas have written.

How can you satisfy some of these wishes without shirking your professional duties, or your need to publish so you won't perish?

Some possibilities:

• First define honestly what you want to write, don't just use the wishes as a way of saying, "I'd rather be doing anything other than writing this grant proposal."

• Realize that you don't need to quit your job and close yourself in a garret should you, for example, decide to write the Great American Novel. There are gradual, possible ways to write outside your professional obligations (though you may discover that writing novels can sometimes be as agonizing as putting grant proposals together). There are a number of famous writer-doctors who both held down full-time jobs and wrote major works of fiction. And if you can manage to do a different sort of writing, you may experience carryover from your "hobby," and find yourself more relaxed and productive as a scientific writer as well. Think of writing of any sort, especially a kind you can enjoy, as building your writing muscle for the writing you have to do.

For example, promise yourself some regular, small bits of time (even 15 minutes a day) for your "fun writing," whatever it is, knowing that producing fiction, essays, or poetry can make you better at other kinds of writing. Consider using some of that time for reading the genre that you'd like to try out.

Know that you don't have to make any sudden, drastic changes in your writing process or product, and that you can modify your style gradually, for example by using the first person, or less formal language, or shorter sentences. Think of these changes as a series of iterations.

If you've been longing to write a single author paper, you don't need to tell your coauthors that you're withdrawing from your joint projects. Start by thinking about what that paper of your own will be, and take your time with it, knowing that you don't need to finish it for a deadline (this in itself will be a liberating experience!).

If you want to write more coauthored papers, begin the process gradually, see if it feels the way you expected it to, and if it does, move over time toward that mode.

Moving toward writing what you want to write will not only strengthen your muscle, it will also increase the quantity of writing you can produce. Once again, behavioral psychology can explain this: when you write in a way you enjoy, you reinforce the behavior of writing in general.

But you may also find yourself better able to tolerate even your formerly unpleasant professional writing, becoming able to separate its "have to" quality from your genuine interest in the material, remembering why you chose the work you do.

Becoming clear about the writing you want to do and how you want it to sound, exploring other genres as well as different kinds of authorship for your professional writing, and separating the fact that a paper is required from your inherent interest in its subject, are all ways to move toward writing more easily, becoming more productive, and owning your work.

23. Writing More, Faster and Better, and Suffering Less

One common problem is having too many writing projects, each of which takes too long to do.

How might you put "more" and "faster" together, when it's already difficult to write slowly?

Consider the possibility that starting a writing task by working at it carefully and painstakingly may not be the most efficient way to get it done. (Note the etymology of the word "painstaking," which seems to include "being in pain," as well as "taking pains" in an attempt to get it right the first time.)

Writing to get it right the first time is like driving a car with the emergency brake on. In order to write more good stuff faster, and suffer less, you need to focus on removing the stalling, obsessing, and nitpicking from your composing process, and think about a different approach.

Writing slowly is often a stop and start operation. But if you pay attention to your own thought processes, and try writing down whatever comes into your head in the course of a few minutes, you'll note that your transcription moves quickly, and sometimes presents you with surprising ideas.

Here's a different way to write that can produce less suffering:

• When you start out, try writing faster, and lowering your standards (this isn't a typo: I really mean "lowering").

• Don't allow yourself to get stuck on one idea, or seduced by working on a very small piece of your prose, for too long. Your aim should be to get as much of a "zero draft" as possible, as quickly as you can. Later on, you can revise to your heart's content, and work toward getting it perfectly right, so long as you realize that you'll only approach, and not achieve, that goal.

I'd guess that some of you are experts on the subject of pain (both professionally and in your writing), know a fair amount about the connection between pain and terror, and understand that being scared makes pain worse. Which is why, when you write faster ("less painstakingly"), your pain will often diminish — or even disappear.

One of the liberating surprises is that writing cures anxiety.

A few more observations:

• Writing slowly tends to increase anxiety.

• Writing quickly tends to diminish anxiety, to get you past it, and into your real work.

• Being over-impressed by your own anxiety isn't useful. If you can't ignore your fright, write about it in the midst of your text (just remember to delete it before you pass the paper on to your coauthor!).

But "writing scared" can be very useful:

Some writing anxiety is quite functional: It provides the edge that you need to leap into a project.

It's also useful to find out that you don't need to wait for your anxiety to go away in order to write.

There is a joke about this problem among therapists who work with academics: that there are more efficient ways than psychoanalysis to cure writer's block (this is what passes for humor in my profession). I'm a clinical psychologist, deeply invested in helping people understand what holds them back in their lives. One of the things I've found over a career working with stuck writers is that pragmatic and behavioral strategies are often of much more use to them than deep therapy: that indeed, writing can cure (some) anxiety.

Writing faster accomplishes several things at once:

• It breaks through the "stuckness" that often occurs when you set out to get it perfect the first time, and allows you to get past first sentence terror.

• It gets a lot of ideas written — and it doesn't matter that not all of them are fit for the final product, or even necessarily correct.

• It encourages thinking, because it more closely matches the speed at which ideas come. Slow writing can make us forget or ignore the ideas that speed through our minds, and these thoughts are a potential source of our richest work.

24. Lowering Your Standards, Producing Better Writing

The title of this post may seem paradoxical, but it isn't. From many years of working with writers (including some who've needed to meet a very high standard) I've discovered that beginning a writing project by trying to get it perfect — or at least excellent — on the first attempt is more likely to produce a poor product, or none at all.

But what I mostly want to talk about isn't the product, but the process by which you create it.

First though, some reminders:

• Perfection is impossible;

• But you do have to set a realistic goal for each of your writing tasks, and do a cost-benefit analysis for each project: how much work to put into a piece of writing, based on how important it is. (Will the grant proposal, if accepted, support your research for years? Is this article just a summary of some work that's interesting, but not groundbreaking? Will a clean draft of your report suffice?)

In other words, not all projects are created equal, or deserve the same amount of effort.

Post number 22 discussed writing quickly vs. writing slowly, and which is likely to produce better results. Its conclusion is surprising: rapid writing tends to make for better writing.

Trying to get your document right the first time confuses the two very different processes of creating and revising. And it slows you down. It's more sensible to follow William G. Perry Jr.'s advice to "First make a mess, then clean it up."

Why can faster, messier writing at the beginning produce a better ultimate product? When you allow your writing to take you where it will (rather than forcing it into predetermined channels prematurely), you are also allowing for thinking, not just reporting.

But perhaps most important is the difference that lowering your standards can make in how you approach your writing. Setting an initial bar too high very often leads to putting off beginning, or keeping at your project, because you've set too formidable a goal. B.F. Skinner called this psychological process "extinguishing behavior" — that is, by setting up a task so that it feels like punishment, you effectively make it harder and harder to approach. Imagine deciding you have to run five miles a day when you've previously run only one: you're likely to give up on running altogether.

One common cause of writer's block is setting unrealistic goals, so that you disappoint yourself each time you don't meet them, and begin to think that writing is just something you're unable to do. So you stop.

Lowering your standards doesn't mean accepting substandard writing; what it does is allow you to relax a bit while you write. This produces a better product: paradoxically, lowering your standards ends up raising them, without your having to terrorize yourself.

The bottom line: lower your standards, start off with a quickly written mess, do several drafts (much easier once you've got something — anything — written), and see what happens.

25. Questions to Ask Yourself About Writing

You may not want to tackle all of these questions at once, but read them, and mull on them over time. (You might even consider writing about them.)

• How much time do you spend writing at a time? How often do you write?

• What are the best circumstances for getting your writing done? The worst?

• Do you write on the computer, use voice recognition software, or write by hand best? Have you experimented with different modes?

• Where do you write? Out of choice, or necessity? Does it work?

• Whom do you write for?

• Do you know in advance what you're going to say, or does it develop as you write?

• Which sorts of writing are hardest for you? Easiest? Do you know why?

• How many projects do you work on at once? How do you organize them? How do you rank their importance? Do you give them equal attention?

• Are you an anxious writer? What makes the anxiety worse? Better?

• What was your best writing experience? The most miserable one? Do you know why?

• Who taught you to write? If no one, how did you learn?

• Who's your best critic? The most considerate one? The worst? Why do you show your work to that person?

• Whom would you most like to write like? Who's the writer you most admire? Why?

• Do you read? Does your reading have any impact on your writing?

• What's the best piece of writing you've done, and how did it get that way?

26. There Are No Personality Transplants

There's an old joke about personality transplants I've used in many lectures: "As of now, there are heart transplants, kidney transplants, gene transplants ... but no one has yet successfully performed a personality transplant." But the reality is serious: all writers are stuck with doing it as who we are, not as some paragon who does it better, or more easily, or neater, or faster, or whatever. What's important about the personality transplant joke is not only that they're not available, but that it captures an important reality.

What follows from this reality? You need to come as you are, and to write that way. Writers often hide the way they really work, because they're embarrassed, and think that other people have neater, more efficient, or clearer methods than they do. Not so. Others have different ways of writing that may or may not resemble yours, and may or may not be efficient, neat, or successful.

Here are a few ways of writing as yourself, and ideas about how you might accomplish it:

The most important thing to do is to stop beating yourself up for being who you are, and figure out how you're going to take full advantage of that reality. For example:

• If you're messy, don't waste time trying compulsive cleanliness. Mess can be positive: a good growing medium for ideas, one that doesn't extort a high entry fee for beginning to write, and that encourages divergent thinking. (Many productive scientists admit to having ideas come as inspirations, and then having to make them sound like they were the outcome of rational thought).

• If you're well organized by nature you can always lay your hands on the relevant data, papers, or drafts, and you're freer of anxiety about how your product is going to turn out. Don't try to be messy, thinking it will make you more creative.

• If you're an anxious writer, work to control and harness your anxiety: realize that pacing up and down your office is part of the process. Teach yourself not to be overly impressed by anxiety, and consider renaming it "adrenaline."

• If you're someone who works on things for a long time, remember that quickness is not next to godliness, fast is often superficial, and simmering is useful. Moreover, if you work on more than one project at a time, slow and steady can be very, very productive.

• If you write at strange hours, know that so do lots of writers (a doctor who's published several popular books writes from 1 to 4 a.m.). If it produces good stuff and you're not destroying your health, your other work, or your family life, go for it. There is no "best time for writing," just your best time (but be sure your proclivity for odd hours isn't a result of procrastination).

Don't fight who you are: It's a losing battle. Look at the ways you write, investigate them, use them if they work, fix them if they don't.

I once read a wonderful obituary of a very productive elderly scientist who hadn't cleaned her desk for the last 30 years of her life (but had continued to write and publish up until the end). How is this possible? Eric Abrahamson and David H. Freedman write about this in <u>A Perfect Mess: the Hidden Benefits of Disorder</u>:

"Though it flies in the face of almost universally accepted wisdom, moderately disorganized people, institutions, and systems frequently turn out to be more efficient, more resilient, more creative, and in general more effective than highly organized ones."

While I wouldn't suggest you apply this proposal to your lab, I think it's very much worth considering when you're thinking about your writing. For many serious writers, writing isn't a neat process, and the benefits of "mess" are clear (if you change the word to "thought compost" you get some idea of why). But if you're a neat writer, and that works well for you, it's equally important that you don't try to turn yourself into a messy one: the message is "come as you are."

How can you find the process that works for who you are, rather than who you'd wish to be? Some visual thinkers put flowcharts up on the wall, because that helps them to see the structure of their papers. Some feel like they can't begin writing until they have not only all the ideas, but also clear outlines of their papers. Others take a more "organic compost" view, and keep adding new ideas (sometimes even ones contradicting earlier ones) and turning the pile over, to see what emerges. And some use the writing itself to draw them toward clarity and coherence.

Most of you probably know how your writing gets done, even if you disparage that process. I'm suggesting that you look hard at the process, not in the abstract, but in its particulars, and if it seems to work, try to figure out how and why, or which parts work better than others. You can decide to change some of the parts that don't work well — just don't think you can turn yourself into a different person.

27. Reentering Your Writing

Here are some strategies for coming back to your writing, and the projects that are waiting for you after there's been an interruption — especially, but not only, in the form of a vacation.

I once had a writing client who announced that because he'd been on vacation, and hadn't written, he felt he had to double his daily production of pages for a while in order to catch up. I told him that this strategy was far more likely to reduce his production, and that punishing himself for having taken some much-needed time off was a good example of negative reinforcement, i.e., that it was far more likely to make him stop writing.

What vacations or time off often produce is new energy, and new ideas for your writing.

Most of our lives include interruptions to work, whether they take the form of chosen interludes, life crises, travel to conferences, other pressing events, or even writing blocks. Here are some ways to reenter your writing that will help get you back in:

• Don't punish yourself the way my client did. The time you didn't spend writing is in the past; you can't rewrite the past. What you need to think about is moving forward. So begin gently, not expecting to be up to your usual speed right away, nor to be able to put in the long stretches you can do when you're in the midst of a major project (a good analogy is returning to an exercise routine).

• Don't tackle the hardest piece of your project first. Aim for smaller goals the first week (otherwise, you may find yourself becoming both frantic and unproductive).

• See if you can cultivate an attitude of positive anticipation, and return to your writing as an interested reader. Reread where you left off, and use that as a place to begin. Sometimes when you give your brain a rest, you'll find that you have new ideas when you reengage with your project, or that you're further along than you thought you were when you last worked on it.

• But if where you left off is incomplete and a mess, accept that, and query the mess: "What was I trying to say here? I can see something lurking in this sentence: can I clarify it? Is this assertion really true? I can see the beginning and the end of this paper; what's missing in the middle?" Think of the mess not as junk, but as an early draft that provides you with something from which to work forward, part of the process of revision.

• An interruption can also be a chance to experiment with some parts of your writing process: think about whether you want to try out a new place, or another time of day than when you've usually written, or a different approach ("I always outline first; what would it be like if I try just writing and seeing what emerges?" or, "I'm focused on what the granting agency will think of this proposal, whether they'll see it as fundable...let me start off instead by just plowing into the exciting research I want to describe, and then I can put it into proposal language when I revise," or, "How can I say what I need to say in this article, while still incorporating the angles my coauthors want?").

28. How Do You Define "Productivity"?

There's a lot of emphasis on "increasing productivity" at most highly ranked research institutions. How it's usually defined is some version of "x publications in y time." Even when productivity seems like just a numbers game, though, it's not only that. Quality matters, clearly, as measured by grant awards, or the status of the journal in which a paper appears.

But there are other ways to define productivity, and some real dangers in limiting its definition.

In "A guide to increased creativity in research: inspiration or perspiration?", Craig Loehle writes,

> There are four requirements for a successful career in science: knowledge, technical skill, communication, and originality or creativity. Many succeed with largely the first three. Those who are meticulous and skilled can make a considerable name by doing the critical experiments that test someone else's ideas or by measuring something more accurately than anyone else. But in such areas of science as biology, anthropology, medicine, and theoretical physics, more creativity is needed because phenomena are complex and multivariate.

He goes on to write about "certain strategies that may promote creativity," and to note that "pressures on scientists today oppose truly creative thinking."

One question I think we need to look at is whether productivity as narrowly defined is, in some ways, inimical to creativity — and if so, how we can think more carefully about expanding its definition. My concern is that too narrow a definition of productivity may penalize those who work on deeper and more complex problems, and encourage a very narrow view of what's worth doing, and what's too risky to work on because it may not pay off.

97

But there are some very different ways to think about productivity.

Consider the *Boston Globe*'s obituary of Judah Folkman, written by Scott Allen. The "p word" wasn't explicitly mentioned, but much of the article was really about it:

> Dr. Judah Folkman, a world famous cancer researcher whose insights led to a whole new field of medicine, knew that his relentless pursuit of ideas could wear people out. For 36 years, sometimes in the face of deep skepticism, the renowned researcher at Children's Hospital Boston stuck by his belief that tumors could be stopped by cutting off the blood supply they need to grow — even when his experiments sometimes fizzled.
>
> "If your idea succeeds, everybody says you're persistent," Dr. Folkman liked to joke. "If it doesn't succeed, you're stubborn."
>
> When Dr. Folkman's research made national headlines in 1972, colleagues accused him of offering people false hope about breakthroughs that had not yet occurred.
>
> In hindsight, Dr. Folkman agreed some of his comments had been premature, but he never gave up trying to prove tumors would shrink if their blood supply dried up. Over time, Dr. Folkman's increasingly impressive results treating cancer in mice wore adversaries down, persuading many of them to join him in looking for compounds that could shut down the formation of blood vessels feeding tumors.
>
> Dr. Folkman did care passionately about ideas and pushing them as far as they could go....

I was left wondering if the interesting aspects of Folkman's attitude (stubbornness, risk-taking, long vision — and brilliance as well) would be permissible in this era of (forgive me) bean-counting.

I've also been dipping into Alan Lightman's book <u>The Discoveries: Great Breakthroughs in 20th-Century Science</u>, a collection of original papers by scientists from Einstein and Bohr to McClintock and Krebs. In Lightman's extraordinary prefaces to each, he investigates and speculates about how discoveries get made.

What the obituary and the book have in common is an implicit (and quite similar) focus on how first-rate (whoops! numbers again...) science gets done.

I'm not so naive as to believe that bean-counting can, or ought, to be totally dispensed with. If I were to ignore the necessity for funding, and the usual measures of who deserves support, or promotion, or better lab space, I would quickly be reminded of them by you and others.

Here are some questions to consider:

• How does the numbers game influence the way you work and write?

• Do you think you'd do more or less research, better or worse, under a different system?

• If the system stays the same (likely), are there ways for you to do some part of your work in a way that you think would please you more?

• How much does the current reward system influence the problems you choose to work on? In what ways?

• Do you have a larger project — perhaps the kind that requires risk, and the sort of "stubbornness" alluded to in Folkman's obituary — that you'd like to work on? Can you? Why or why not?

• Can you pay attention to "the beans" while pursuing the research that you're most interested in and excited by? If so, how?

- How does coauthorship affect your productivity (if it does)?

- If you could do any research you wanted, what would it be?

29. Why Do You Write? Whom Do You Write For?

These sound like simple questions, but they're not. You may immediately respond, "I'm writing for the journal I hope will publish this paper," or "for the doctors/researchers out there who need to know about what we've discovered." Or, in another realm, "for the committee that decides on my promotion/tenure/salary," or "for NIH reviewers, so I can get a grant."

All of the above answers are reasonable, and instrumental, and that's OK. But this approach to writing has repercussions for both the finished product, and for you. If you write exclusively for others, it can cause problems. Being a brilliant poseur isn't the same as being deeply invested in what you've written. What to do about this? Sneak more of your own thoughts in, even if they're disguised?

Consider an alternative: begin by writing for yourself.

One way to do this is to make the earliest draft of your paper say everything you want to say, including thoughts the ultimate reader might hate or disagree strongly with, stuff you're not sure of but find provocative, new ideas, and other things you're pretty sure are true, but that may not be welcomed by journal editors or grant reviewers.

Writing needs to satisfy the instrumental, personal, and moral needs of the writer, not only the requirements of the readers. What is most important is that you know where the truth lies, and that you don't cross a line that will leave you feeling uneasy. And remember that writing is not only "writing up" results, but a mode of discovery. You can't know exactly what your conclusion will be until you get there.

Then move on from that first draft: keep the original for yourself, and tailor what you write in the finished piece to the necessary external requirements. Sometimes writing what you know your audience wants to hear will be necessary. But it will feel different if you've already written for yourself.

30. Hitting the Wall and Writing Paranoia

Hitting the wall is a common problem for writers. It can take many different forms, from "I have nothing to say...nothing worth saying, anyway," to "no one is going to want to read/publish/fund this," to "I've used up every decent idea I've ever had."

These feelings have something in common. They're all a bit on the paranoid side, and can make you think your glass is empty, not just half-empty. If you believe them, they're convincing excuses for not writing.

I don't mean that we consciously invent these excuses so we don't have to write. I think that when we're in this sort of mindset, we really believe they're true.

The causes? Sometimes it's as simple as sheer exhaustion, endemic in the scientific/medical world. This sort of weariness comes along with depression, and with the temptation to echo Melville's Bartleby's stance of "I'd rather not." We give ourselves this signal when we're depleted — fortunately, it's usually temporary.

Sometimes stopping writing has to do with the project itself, for example when the results of a study aren't yet making sense. You may feel that you're in over your head, or not getting enough support. And sometimes you just give in to your more paranoid fantasies.

When the project involves coauthors you may feel you don't own the work, that it isn't really yours; or you're not comfortable with the way the work has been divided up. Or the group may just be incompatible in how you approach problems, or in your strengths and weaknesses.

You may stop dead in your tracks because you're afraid of the feedback you'll get from your readers (this is especially true if you've had too many harsh responses in the past — or a horrific one from someone who still judges your work).

How might you deal with these hard and complicated feelings?

I've written about ownership before, but I think it's an issue that's particularly important if you've hit the wall.

Ask yourself questions like these:

• When did you stop writing? What was happening then?

• Do you think your coming to a halt had to do with the work, or with your life?

• Has hitting the wall with your writing happened before, and if so, how did you get over it?

To explore a bit further:

• Try writing <u>about</u> the paper, asking yourself why you might be stuck. Can you remember being interested in the topic earlier, and if so, what's changed?
• Are you so frightened of the response you'll get to your draft that you'd rather stop dead? (N.b.: Don't submit your earliest draft to other people's scrutiny.) Consider asking someone who won't judge you, but whose judgment you respect, to read your draft, and just to tell you if it's OK for now…<u>not</u> to criticize it, but to let you know that reading it has caused them neither indigestion nor heart failure.

If you want any feedback at all, ask your reader specific questions like these:

• Can you tell me what you think I'm saying?

• Could you follow my argument? Did it make sense to you?

• What was missing?

Perhaps most important is to remember is that all writing goes through cycles, and that even the bad times and drafts are part of the process. You will write again.

31. Other People and Your Writing

Begin by asking yourself some questions:

• What roles do other people play with respect to your writing? (Try making a list.)

• When and where did you learn to write? What messages did you receive as part of that instruction? Did anyone ever teach you to write? Did Ms. or Mr. Horror, your elementary school teacher, present writing as nothing more than rules for grammar and the five-paragraph essay?

• Have there been times when you found writing pleasurable? If so, what do you remember about those experiences?

• Who's helped you with your writing?

• Who now plays a role in your writing? Is it positive, negative, or neutral?

• If you could have anyone to help you with writing, what sort of help would you wish for, and with what sort of a helper?

• Who reads your drafts? Is s/he helpful? If so, how? If not, why not? What might you do about it?

Now that you've interrogated yourself, let's explore the writing relationships you might have with critics, supporters, and anonymous readers. (I'll save the nastier creatures for the end.)

The Critics are always there; sometimes they're you. In her splendid essay "The watcher at the gates," Gail Godwin talks about the critic who lives inside every writer's head. Read it, if you haven't already: you'll likely find parts of yourself there. We can't often change other people's ways of criticizing, but we can work on our own, and the internal critics are very often the ones who stop you from writing.

The Supporters: How might you find them and use them, if you don't already have them? Use the grapevine to find out whom other colleagues of yours have found helpful. Consider a writing group of peers. Seek help from good colleagues, and see which of them are best at helping you. Figure out what you want them to do for you: read your work? Help you stick to your deadlines? Tell you that you're not the brain-dead person you sometimes feel like? Share strategies they've used to keep their own writing moving? Also tell family and friends what you need from them that will help you get your writing done.

The Anonymous Readers who decide if you win the lottery (a proposal funded, an article accepted, a book published): they are often figments of your imagination, since you usually don't know who they are. You can't change them, and praying to them probably won't work, but you can change the pictures of them you carry around (most are probably not person-eating ogres), and you can work on your own responses to their feedback.

Now let's take a look at the negative side of writing "help":

• If the person who responds to your writing makes you feel that both it, and you, have been trashed, or if their criticism shuts off your writing for long periods of time, what might you be able to do about it?

• Try not to show your earliest drafts to anyone (the writing belongs to you, and you get to choose who sees it, and when).

• Don't talk about your thoughts for a piece before you write them down, both because some of them might evaporate if you do, and because this isn't the time to get feedback.

• If a critic is interfering with your writing, assume first that s/he's really trying to be helpful. If so, you might be able to suggest gently some ways s/he might terrify you less, and help you more. But if s/he's known to be devastating to other writers as well (or even just to you) try to switch readers, or, if you are stuck with this one, find someone else who can be helpful, and restore some balance.

Or, you could always create your own reader in your head: maybe the high school teacher who told you you could be a writer when you grew up.

Bibliography

Abrahamson, Eric and David H. Freedman. 2007. <u>A Perfect Mess: The Hidden Benefits of Disorder</u>. Little Brown and Company.

Allen, David. 2001. <u>Getting Things Done: The Art of Stress-Free Productivity</u>. Viking Penguin.

Allen, Scott. 2008. Judah Folkman, cancer's innovative enemy, dies at 74. *The Boston Globe*, January 16, 2008. Metro Section p. A1.

Bamforth, Iain, M.D., editor. 2003. <u>The Body in the Library: A Literary Anthology of Modern Medicine</u>. Verso.

Bernard, Andre, editor. 1990. <u>Rotten Rejections: a Literary Companion</u>. Pushcart Press.

Bolker, Joan L. 1998. <u>Writing Your Dissertation in 15 Minutes a Day: A Guide to Starting, Revising, and Finishing Your Doctoral Thesis</u>. Owl Books/Henry Holt and Company.

Eisenberg, Anne. 1994. E-mail and the new epistolary age. *Scientific American* 270(4):128.

Elbow, Peter. 1973. <u>Writing Without Teachers</u>. Oxford University Press.

Godwin, Gail. 1977. The watcher at the gates. *The New York Times Sunday Book Review*, 31.

Goldberg, Natalie. 2005. <u>Writing Down the Bones: Freeing the Writer Within</u>. Shambhala Publications Inc.

Gopen, George and Judith Swann. 1990. The science of scientific writing. *American Scientist* 78:550-8.

Kidder, Tracy. 2003. <u>Mountains Beyond Mountains: Healing the World: The Quest of Dr. Paul Farmer</u>. Random House Inc.

LaCombe, Michael A., M.D., editor. 1995. <u>On Being a Doctor: Voices of Physicians and Patients</u>. American College of Physicians.

Lightman, Alan. 2006. <u>The Discoveries: Great Breakthroughs in 20th-Century Science, Including the Original Papers</u>. Vintage Books.

Loehle, Craig. 1990. A guide to increased creativity in research: inspiration or perspiration? *BioScience* 40(2):123-129.

McPhee, John. 1984. <u>Heirs of General Practice</u>. Farrar, Straus and Giroux.

Roueché, Berton. 1947. <u>The Medical Detectives</u>. Penguin.

Thomas, Lewis, M.D. 1992. <u>The Fragile Species</u>. Touchstone/Simon and Schuster.

Cover design: Paul Mason
Technical and editorial assistance: Jessica and Ethan Bolker

www.ingramcontent.com/pod-product-compliance
Lightning Source LLC
Chambersburg PA
CBHW022025170526
45157CB00003B/1356